If anybody can worm their way into the hearts of the World's sportsmen, gardeners, and econologists — you can. Good Luck. To George Sroda—Godfather to our little friends.

       Jerry Chiappetta
       Michigan Outdoors TV

Not only a friend to the gardener and fisherman ... but a friend as well to the lowly worm. Sroda cares—and the worms know it!

       Mark Zelich
       Director of News and Sports
       WSAU TV, Wausau

Take my word for it, if you want to know anything about nightcrawlers for fishin', ask George Sroda: The Worm Czar.

       Harry Bonner
       Fishing Editor
       SEA Boating/Fishing Almanac:
       Mexico to Alaska

I have your letter of the 17th and think the book is a smart way to go. How about the following? "Seemingly a wormy put-on ... but every word is gospel. No one knows the worm better than George Sroda."

George, you may chop up the above quote as you wish ... it will still be a can of worms.

       Dale Lemonds
       Producer/Director
       WTMJ TV, Milwaukee

There is no doubt in my mind that nowhere in the world is there anyone more knowledgeable about Earthworms than the Worm Czar, George Sroda. His research into the care, keeping, feeding and various diseases of Nightcrawlers is truly amazing.

       Jim Wrolstad
       Senior Editor
       Fishing Facts Magazine

I have received a great deal of help for the many questions I asked you in regard to worms, which is my business. I can only say, you are a true Worm Expert and recommend your book "Facts About Nightcrawlers" to all.

                         Karen Pirk
                         Karen's Bait
                         King, Wisconsin

I'm sure everyone who likes to fish will be just crawling to get their hands on a copy of your worm book. It'll be well worth it too, because if anyone knows the ins and outs of nightcrawlers, it's a fact that you do.

                         Jim Harp
                         Outdoor Editor
                         The Post-Crescent

George Sroda, the Czar of the Underworld at his best. Facts About Nightcrawlers is the most comprehensive informative and practical guide yet.

A must for the angler and bait dealer ... and the home gardener and professional horticulturist too. The Oligochaetes never had it so good.

                         Tom Boario
                         Sports Editor
                         Waupaca County Post

Congratulations! It is quite a book ... I like a book like this which is practical, easy to read and inspirational. Considering I have now special interest in worms, I was surprised to find how interesting it was to me.

                         Anna Glidden, Poet
                         "The Breeze Changes"
                         Palo Alto, California

No Angle Left Unturned:

FACTS ABOUT NIGHTCRAWLERS and REDWORMS

by
George Sroda

Copyright 1975 by George Sroda

First printing 1975

Second printing 1975

Third printing 1982

Fourth printing 1984

Fifth printing 1988

Sixth printing 1991

Seventh printing 1995

All rights reserved. No part of this book may be reproduced or utilized in any form or by any means, electrical or mechanical, including photocopying, recording, or by any information storage and retrieval system, without permission in writing from the publisher.

ISBN 0-9604486-0-8

ACKNOWLEDGEMENTS

I would like to offer my thanks to all the people who have helped me in my work and encouraged me to write this book. Among them are: Janet Hughes, B.M.S. Inc., Mark Zelich, Jerry Chiappetta, Harry Bonner, Dale Lemonds, *Fishing Facts* Magazine, Karen's Bait Shop of King, Wisconsin, Jim Harp, and Bill Hoeft.

I would like to offer special thanks to the staff at the University of Wisconsin-Stevens Point Writing Lab for their help with organizing, writing and editing my text. I want to acknowledge the valuable advice and assistance given to me by Tom Buchholz, Donna Nelson, and Mary Croft.

I would also like to thank my loving wife, Susan, without whose devotion and understanding my life and my work would be meaningless.

George Sroda
Amherst Jct., Wisconsin, 1975

NO ANGLE LEFT UNTURNED:

FACTS ABOUT NIGHTCRAWLERS

Attention Universities, School Clubs, etc.

NEW . . For your review . . VHS or BETA. An eighteen minute tape by George Sroda, International Worm Czar (worm expert) who rules over two million earthworms in the largest research laboratory in the world. George feeds worms . . talks to worms . . raises worms and eats worms. His star, "Herman the Worm" is the most photographed and talked about worm in the world . . a worm that paints modern abstract designs . . a worm that dribbles and makes baskets with his small basketball . . a worm that has wormed his way into the hearts of over 150 million viewers on the small screen. George and Herman have been guests on over 22 National TV shows in the United States, from "Coast to Coast" and also aired on satellite . . Dublin, Ireland. A story that is factual, educational and humorous.

A Biology Teacher from Germany visits the Worm Czar's Laboratory for worm knowledge to take back to his classes in Germany.

A few photos of some of the 22 National TV Talk Shows George and Herman have appeared on.

Some people think that a worm is a squirmy thing that you buy in a box, put on a hook, and use to catch fish. But there are other uses for earthworms. (1) Worms for bait to catch fish. (2) Worms for food. Those which contain earth are first stripped and the earth is removed. Then they are cooked in hot water with other ingredients for hours. (3) Worms for medicine. Worms are ground with other herbs. (4) Worms freeze dried for fish food. (5) Worms for plants. They aerate the soil and their castings add fertilizer to the soil. (6) Worms for pop art. Dipped in nontoxic paint, they become living paintbrushes.

FOREWORD

Most people consider the dog to be man's best friend. I, George Sroda, known as the "Worm Czar of the Nation," think man's best friend is the earthworm.

Worms level hills and fill in valleys. They work up hard soil like a plow, permitting air and water to reach plant roots. They chop up rocks in their powerful gizzards, adding fertility to the soil as they work. Considering all the demand for preserving ecology, I say that worms are the answer to those who fear chemical fertilizers and their effects on soil.

Earthworm farming is a big business and it's getting bigger each year. It is not uncommon for large earthworm breeders, mostly in the South, to have an income of over $40,000 a year. The biggest market for earthworms is the fisherman. Fishing license sales in 1972 were 26,022,547. Money spent was more than $107,000,000 on licenses alone. Fishing is getting to be our number one sport. When people have nothing else to do, they go fishing. When they have long weekends, long vacations, and layoffs, what do they do? They go fishing. There will never be an oversupply of earthworms for fishermen.

Very little information has been obtained and released on earthworms recently, except for the work done at the Laboratory. I think this research is a valuable source that should be given to the public today. I also think that the most valuable animal is the earthworm. As an oligochaetelogist, one who studies worms, I have put much effort into this publication, and highly recommend it to all 50,000,000 fishing enthusiasts, of whom over 10,000,000 are women. I recommend it to anyone interested in better methods of holding and conditioning earthworms, especially the great native nightcrawler.

I do not sell worms. I spend hours and hours each day working with over 2,000,000 breeding worms, baby worms, and worm-egg capsules. I have employed the best and latest methods in my research. I've had years of experience in merchandising, and have included a chapter on merchandising earthworms.

I believe that too many worms die because of improper care. There is a great need for information to assist worm handlers, especially because of the high cost of worms.

Because of my earthworm experiments, knowledge and reputation, I have appeared as a guest on many local and national television shows, such as "Johnny Carson's Tonight Show" . . "Mike Douglas Show" . . "What's My Line? To Tell The Truth" . . "Bill Baker Show" . . "Michigan Outdoors" . . and "Wisconsin Outdoors" . . I have been on national radio, and have been featured in a number of state and national newspapers and national fishing magazines, including *NATIONAL FISHING FACTS*. I haven't stopped here and I'm not looking to retirement at this time. I have many years ahead of me in this work and I will continue to find more and better information to bring you from my research. Earthworms are my life's work and I love it.

I have been encouraged to write this book by many good friends—sports writers, my past and present television hosts, radio reporters, operators of small farms, bait dealers and fishermen. My knowledge of photography has also played an important part in this publication. This is a book with many illustrations and pictures that I have taken while at work with my worms. This book is written in simple, easy-to-read language and is recommended for all — from the fishing enthusiast to the worm breeder who wishes to have better, healthier worms. This is a book for both sexes and all ages. It will be especially interesting to the retired person who wishes to supplement his income through raising worms.

This book answers many questions sent in to me by people throughout the United States, asking my advice in regard to earthworms. Chapters are designed so that they can be read randomly. For example, a worm breeder might find Chapters 2, 3, 4 and 8 very interesting. A fisherman might want to read Chapter 5 more than once or everytime he plans a fishing trip. Gardeners will be attracted to Chapter 7. On the other hand, all of the book is meant for all of the people. So, here are the Facts from the Nation's Worm Czar.

Thank you and good luck,
George Sroda, "Worm Czar"

TABLE OF CONTENTS

Foreword 11

Chapter 1
WHY THE GREAT INTEREST IN
THE NIGHTCRAWLER

Unusual Uses for Nightcrawlers 23
Earthworms as Food and Medicine 30
Breeding Nightcrawlers 30
Other Useful Facts about Nightcrawlers 33

Chapter 2
HARVESTING THE NIGHTCRAWLER

Equipment 39
Picking up Nightcrawlers 45

Chapter 3
HOLDING NIGHTCRAWLERS

Building a Master Worm Box 53
Bedding 58
Watering 59
Controlling Temperature 61
How You Can Cool the Bedding 63
Amount of Bedding 64
Feeding Time 64
Keeping Your Worms Happy 68

Chapter 4
MANAGEMENT OF WORMS DURING
THE HOLDING PERIOD

Mixing Worms 76
Cautions 76

Chapter 5
TRANSPORTING WORMS DURING FISHING TRIPS
On the Road 84
Out Fishing 85
Baiting Your Hook 86
Ice Fishing 87

Chapter 6
THE REDWORM
Breeding Redworms 95
Getting Rid of Flies and Mites 96
Sex Life of Redworms 98

Chapter 7
WORMS AND SOIL BUILDING
How to Plant Worms in the Soil 107
Earthworms for Monitoring Soils 108
Good Worms Make Good Gardeners 108
Attention Ladies...Free Organic Potting Soil 115
   (Composting Project)

Chapter 8
MERCHANDISING WORMS
Don't Let Your Worms Lose Their Cool 120
Selling Worms 121

CONCLUSION

IN A NUTSHELL
Tips for Fishermen 130
Tips for Worm Breeders 131
Dear Anglie 132
Additional Worm News 134
George Sroda: Central Wisconsin's Unique
   National Celebrity and Promoter
   Extraordinaire 138

SOME OF THE STARS AND CELEBRITIES
I HAVE APPEARED WITH ...

Johnny Carson . . Mike Douglas . . Garry Moore . . David Letterman . . Priscilla Presley . . Burgess Meredith . . Jim Stafford . . Bill Rafferty . . Terry Meeuwsen . . Soupy Sales . . John Denver . . Flip Wilson . . David Brenner . . Schecke Green . . Mel Tillis . . Rita Moreno . . David Niven . . Pointer Sisters . . Anthony Newly . . Ed Asner . . Fred Gressith . . Jerry Chiappetta . . Marsha Mason . . Dr. Joyce Brothers . . Alan Thicke . . Bobby Benson . . Olivia Brown . . Kitty Carlyle . . Peggy Cass . . Gene Schallet . . Bill Cullen . . Melba Tolliver . . Leonard Harrison . . Arlene Francis . . Sarah Purcell . . Skip Stevenson . . Ed McMahon . . Joe Flynn . . Brian Keith . . .

There's no doubt about it, George and his worms have wiggled their way into the hearts of over 150 million of America's television viewers during more than 500 appearances on the small screen.

They have taken the stage by storm on the "Tonight Show" . . "Mike Douglas Show" . . "Amazing Animals" . . "Real People" . . "Winners TV Show" . . "To Tell The Truth" . . "What's My Line" . . "Kid's World" . . "PM Magazine" . . The "David Letterman Show" . . "Late Night" . . "Sports Unlimited" . . "Michigan Outdoors" . . "Outdoor Wisconsin" . . "Kelly & Co." . . "Morning Scene" . . "Sports Afield" . . "Late Show" , Dublin, Ireland . . As well as many other television & radio shows . . . .

George, The Worm Czar and his TV star and celebrity Herman The Worm

TV Star Celebrity Herman the Worm

Wormed their way into the Whitehouse.

Wormed their way into Hollywood.

Wormed their way into New York's Mayor's Office, New York.

Wormed their way into Fourteen National TV Shows.

Wormed their way into International TV Talk Shows.

Wormed their way into a Number of Newspapers and State and National Magazines.

Wormed their way into the Fishing Hall of Fame, Hayward, WI.

Wormed their way into the Hearts of Millions of TV Viewers.

TV celebrity "HERMAN THE WORM"

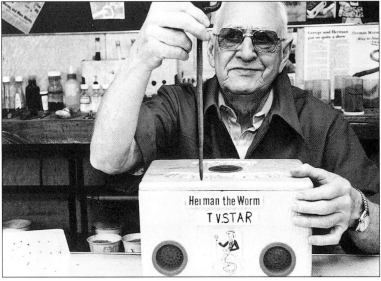

George and Herman The Worm have wiggled their way into the hearts of over 150 million world wide television viewers on the small screen.

Chapter 1

WHY THE GREAT INTEREST IN THE NIGHTCRAWLER

I don't play Golf . . I don't Bowl, I am fighting to save the earthworm, man's best friend.

In this book I am going to discuss the number one earthworm, the nightcrawler, a segmented worm in the annelid (Lumbricus Terrestris) class. Sometimes this worm is referred to as the crawler or Canadian nightcrawler, the dew worm, the grass worm or the rivershore walker. Nightcrawlers are usually found in grassy areas, along river banks and where there is plenty of vegetation, plenty of moisture and food for these worms to survive. They are also found on good lawns, golf courses, parks, where organic fertilizer is used in place of commercial fertilizer and where grassy areas are heavily watered. Here is where you will find the native nightcrawler.

Unusual Uses For Nightcrawlers
Most people think the nightcrawler is a squirmy thing you buy in a box and use for fishing bait alone, but there are other uses for

The Night crawler: Lumbricus Terrestris.

I have a letter, in my files, where a party paid as high as $5.00 per dozen for crawlers.

this important worm. For value to man, as I understand, it is used for food. The Chinese, they say, like their worms fried. In China, processed worms are used as Chinese medicine, mixed with herbs. In Hong Kong and China Mainland, there are many kinds of earthworms used for human food. One of these is called rice worms - sold in freeze condition . . In Hong Kong you can buy live rice worms and cook them with egg white - very delicious and healthy - high protein. Also in China one can buy silk worms, fry them - as human food. Sand worms . . boiled and dried with small pieces of lean pork - as soup. Sand worms is for strengthening the lungs and to stop coughs. Dried and boiled earthworms, mixed with other herbs, with the patient drinking the herb tea - to release the blocking of the arteries. One farm in Taiwan, a farmer breeds millions of earthworms and uses the castings as fertilizer, growing large and healthy mushrooms.

No bone or gristle . . one can eat the entire worm . . when using earthworms as bait, you can't lose . . If you don't catch a fish you can always eat the bait.

Pet food companies freeze-dry worms, manufacturing or processing them while still

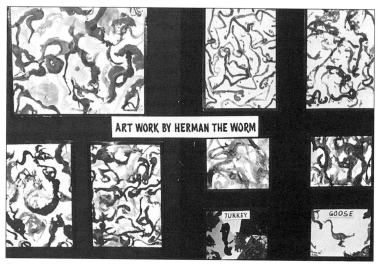

Pop Art. Some time ago, creative artists experimented with Chimpanzees, giving them paints, paper, and brushes. The chimps seemed to enjoy doodling with the paint and turned out some interesting abstract designs. However, I think I topped all that with my earthworms. My earthworms became living paintbrushes. I just furnished them with nontoxic paint and paper, and they created pop art.

alive then dehydrating the ground-up worms into small cubes, about the size of dice. These small cubes are fed to fish.

Earthworms are also used by photographers. The cameraman will use a good, active worm to attract a bird, to get the bird closer to the camera lens or to the area where he is shooting. These pictures appear in outdoor magazines and newspapers. Pop art uses worms to create abstract paintings. The worms are dipped into cans of paint and allowed to wiggle and squirm around on the canvas, producing abstract masterpieces. These have been known to sell for as much as $1500.

Nightcrawlers make good pets. They're clean. They're not harmful. All they need is a little care, proper moisture, proper bedding, proper temperature — and they will stay put, keeping busy around the clock.

Dirt farms use worms, as well as trout farms and fish farms.

Earthworms, especially nightcrawlers, are also used in classroom study throughout the United States in biology classes. The crawler is a large worm and easy to work with.

I have found another market for nightcrawlers: the homeowner who uses worms as feed for fish. Recently, I received a call from a lady in the East. This was during the wintertime when worms were very scarce. She said, "I understand you have worms there. Would you sell me some? I have a certain fish in my aquarium that lives on worms alone. If I don't get some worms for him, he's going to die." Well, I passed some information on to this lady and added that if she wanted to be sure of a year round supply, to go ahead and raise her own worms. I described the right type of bedding to use, the right type of food, and conditions so she could keep worms year round.

I also received a letter not long ago from a lady in California. This lady was writing a new recipe book, using insects as ingredients, and she wanted to know if I had any information on how to use earthworms as ingredients in foods. I didn't have any information then, but thought her request was interesting. Not too long after, I read an article about a young gal in the East who won the National Cooking Trophy by baking some cookies with earthworms. Who knows? With the predicted food shortage, someday you and I may be eating earthworms in various dishes. I'm not kidding! Nightcrawlers are about 70% protein, which is a primary source of energy.

"Earthworms are nutritious and can be delicious. They consist of about seventy percent protein and are rich in minerals — also entirely edible, with no bone or gristle. Processed earthworms can be mixed into cookies, cakes and salads.

"Earthworms as Food and Medicine"

Research Shows: 75% increase in pasture production following the introduction of earthworms. Worms cure chicken disease . . Paralyzed chickens refusing to eat . . All that was done was to feed them earthworms . . They gobbled up the worms . . In three days they were completely cured and were running about . . worms are rich in vitamin "D". "Earthworm Fluids": Chinese chemists distilled a batch of worms in alcohol and created an inexpensive "Earthworm fluid" that can lower fevers, smooth winkles and add nutrition to drinks. Can also be mixed in alcoholic drinks, soft drinks and even cakes.

It is claimed large earthworms could be used in the place of other animals such as dogs, monkeys and rats in the research pertaining to man's health.

Worms, freeze dried are fed to tropical fish, snakes and birds.

Breeding Nightcrawlers

I've been asked about raising nightcrawlers to produce young, in confinement the year round, as is done with the prolific redworm. My answer is this: they will produce in confinement with a properly managed

Mail From All Over The World!

Yes! George Sroda, the Worlds Worm Czar receives cartons of mail from all over the World asking advice about earthworms.

program, but not to the point where it will become profitable. One of the reasons, I think, is that the nightcrawler's habits are a little bit different from the fast breeding redworm. The nightcrawler burrows down deeper into the soil. Some go as deep as eight, ten or twelve feet. They usually lay their eggs in these burrows. On the other hand, the redworm usually lays it eggs near the top six inches of its soil or bedding. So if you think you're going into confinement growing, raising and breeding of nightcrawlers, you'll be very disappointed. You can hold, condition, and grow nightcrawlers indoors, but not breed them very well. The nightcrawler's size will increase, and you'll have a lot better worm than you will by harvesting it outside. Crawlers harvested outside in the spring, after hibernating all winter, are dehydrated to a certain degree.

During the winter you can increase the weight and size of nightcrawlers tremendously by putting them in the right bedding, moisture and controlled temperature overnight. Let me warn you that you can also lose this weight and size overnight by having the improper temperature and moisture. You can grow them all the way from ten to sixteen inches in confinement and that's a mighty nice worm. And then, when the weather is warm and the nightcrawler burrows down into the soil where it's cool and damp (July and August in the Midwest), you'll have earthworms.

Fishermen are especially interested in keeping nightcrawlers over winter. I've had many fishermen come to me and say, "George, if I could get a nightcrawler during the wintertime for ice fishing, boy, could I catch some nice, big fish." So, my answer is that you too can condition them. You can improve their size and activity in confinement, and they will produce some eggs.

Other Useful Facts About Nightcrawlers

If I were harvesting nightcrawlers in the spring to hold for the next six or more months, I would not take the first crawlers that come up to mate and feed in the early spring. I'll tell you why. They have been hibernating all winter and they are weak and dormant. It takes much activity on the crawler's part before he gets hardened and toughened up. And the more you handle these first crawlers, the more chance you have of injuring them because they're soft, tender and weak.

The native nightcrawler has about 150 segments. Injure one of these segments while you're harvesting it or you are handling it and you're going to be in trouble. You know what happens when you injure a segment? The nightcrawler will get sick, start to die and the dying worm will turn into gas and the gas can kill the healthy worms in your bed.

I would wait about six weeks or more, until the crawlers became accustomed to the season. Then I'd harvest them and hold them for the following months.

Nightcrawlers will attain a fishing size in about four to six months. They will mature in about ten to twelve months. The second year when they have the band or collar, they are at an age to mate and produce young. Nightcrawlers can usually grow about an inch or more a year for up to six or seven years.

As you can see, there is great interest in nightcrawlers from people of many walks of life. I think the nightcrawler is not only the most valuable animal in the world, but also one of the strongest for its size. It is claimed that a crawler weighing one-thirtieth of an ounce can move a stone weighing almost two ounces. He's a pretty remarkable fellow. That's why I find him so fascinating to raise and harvest.

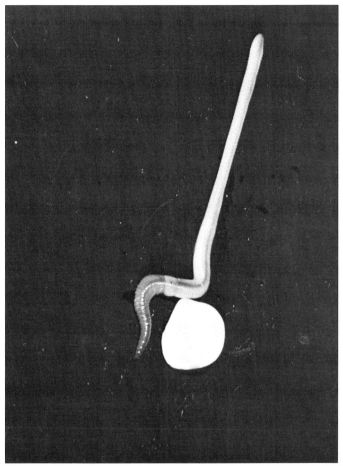

A crawler is known to be the strongest animal in the world for its size. Here we have one of my pet worms, Henry, in the act of moving a good size stone.

Good Worms Make Good Gardeners
I think that vegetables grown in this type of soil, heavily infested with earthworms as the main source of fertilizer, will contain more food value.

Chapter 2

HARVESTING THE NIGHTCRAWLER

When winter is over, and spring arrives at last and you are dreaming of vegetables, flowers and greener grass . . I have an answer that will make you wiser . . always enrich your soil with worm fertilizer.

Equipment

To collect nightcrawlers, you need proper equipment. First you need a good light. You've got to be able to see these worms. It's dark. You're picking them up in the evening, probably around midnight. They're up to mate or to migrate and you want to harvest a nice number of them. So what do you use for light? Instead of a flashlight, which you'd have to hold in one hand, I suggest a light that fastens around your head with a headband, like a miner's lamp. This will free both hands. A miner's lamp can be moved in any direction to focus on the area where the worms are. Then you can hold a container in one hand, and pick up the worms with the other. Most lights come with a white lens, which has a tendency to send the worms down. (Worms are very sensitive to white light.) So I suggest that you tape a piece of red, plastic cellophane over the lens

Harvesting nightcrawlers at night with the "Hands Free Light." The lantern battery power source is fastened to the waist and the adjustable light is on a headband. I cover the light with red cellophane paper.

so you have a red light instead of a white light. Next, you need a container in which to put the worms you are harvesting. Too many people use a tin can with no bedding in it. They overpack or overstock the container, and by the time they get the worms back home, the worms are damaged or overheated. These people wonder why their worms die.

For the proper container, I recommend the Magic Worm Farm Container that I developed. It has many tiny holes for aeration. It has a steel handle for carrying ease and it is large enough to accommodate many worms temporarily. The proper container must also have proper worm bedding which I will discuss in the next chapter.

Be sure that you wear soft-soled shoes, rubbers over your shoes, or a pair of tennis shoes. Or, if you'd rather, go bare-footed. A heavy step from a hard-soled shoe will often send nightcrawlers down. They're very sensitive to sound. They do not have senses like you and me, but they do have a group of cells that are sensitive to taste, touch, and light. Even though earthworms have no ears, they are very sensitive to vibrations.

MAGIC WORM RANCH®

Measuring 14"×20"×7", the Magic Worm Ranch can hold hundreds of fresh, tough, vigorous worms. Complete with 4½ lbs. of Magic Worm Bedding and a 12 oz. bag of Magic Worm Food.

MAGIC WORM FARM®

The 8 5/8"×12"×7 3/8" Magic Worm Farm comes complete with a 1½ lb. bag of Magic Worm Bedding, a 4 oz. bag of Magic Worm Food and a steel handle. It's everything needed to store and carry dozens of fresh, tough, vigorous worms.

I've done considerable research, with millions of worms, in the Worm Laboratory in regard to earthworm containers and recommend these two quality products.

I have often been asked if I believe in using electricity or chemical solutions to drive night crawlers out of the ground. These methods will work, but I don't recommend them. For two reasons: one, the electric rod has killed a number of people and two, the electric method forces the nightcrawler out of its burrow too quickly. So fast, in fact, that in the process the crawler's crop can rupture. This method can also injure other segments and cause death, especially with baby crawlers. As for chemical solutions, they kill too many baby worms while they are still in the soil. Chemicals also pollute the soil and change its characteristics. And incidentally, chemicals destroy the natural odor of nightcrawlers, so important to catching fish. You can attract them to the surface by driving a stick into the ground and striking it several times to cause vibrations. This will sometimes drive them out of their burrows. But do not risk disturbing them, once they are up, by wearing hard-soled shoes.

Now you're ready to harvest nightcrawlers. You've got the container, you've got the proper light, and your hand is free. Here's one more thing. Have a piece of cloth in your pocket somewhere, so you can keep drying your hands. You see, in the spring, these worms are

Harvesting crawlers is a technique. Grasp the crawler below its collar and do not pull when it resists. After a second or two it will relax and you can pull it from it burrow.

migrating, they're mating, and they have a slimy solution on them. The more worms you catch, the more slippery your hands get. You reach a point where you cannot grasp the nightcrawlers properly or you cannot hold them because they slip away from you. If you pinch too hard, you'll injure them. You'll rupture the crop and the worms will die. So, keep your hands dry by wiping them on a cloth as you are harvesting worms.

When you get a fairly decent number of worms in the container, I would suggest that you empty them into a larger worm container. Then you can go back out and get more.

If the nightcrawlers are not coming up to the surface in the evening, it may be due to dry conditions in your area. So sprinkle your lawn heavily all afternoon. That evening they should come up for air and to mate, and you can harvest them then. (Usually they will not appear above the surface if it's under forty degrees or over eighty or during daylight.)

Picking Up Nightcrawlers

Try to grasp them just below the collar, the raised portion of the nightcrawler, located one-third the distance from the head. Get them just below that collar, and give them a slight tug if they don't come.

Nightcrawlers are mostly muscles and nerves. They have two kinds of muscles. One lies along the length of the body, and the other circles the body. When they spread their tails, it is impossible to pull them out without injuring them. So don't pull too hard if they don't come. Just hold on and in a second or so, they'll relax and you can gently pull them out of their burrows. If they do not come, forget about them. Let them go back down. You'll probably get them the next night.

This is one way to get your supply of nightcrawlers. If you don't want to find your own nightcrawlers, or if you have a sore back, you can go to a bait dealer to purchase your nightcrawlers. If you purchase nightcrawlers harvested by youngsters, be sure that the crawlers are in excellent condition and not injured. Injured worms can cause a number of problems in your worm holding boxes.

Naturally, the worms will cost you practically nothing if you go out there and do a little bending and gathering for yourself. Then you can have the satisfaction of picking your own nightcrawlers. But remember, in this phase, many nightcrawlers are damaged or injured. Wait for the right time and be careful.

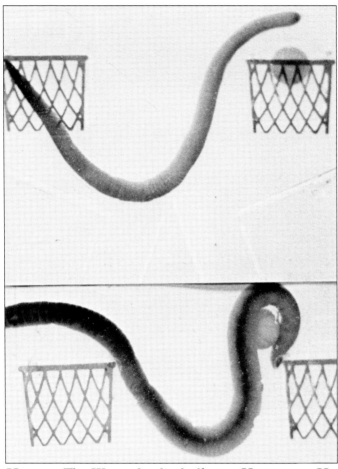

Herman The Worm, basketball star. He turns . . He squirms . . he wiggles and he dribbles making a basket in one. The only worm basketball star in the world.

Chapter 3

HOLDING NIGHTCRAWLERS

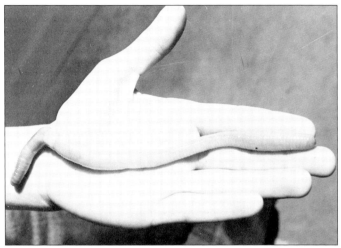

There are various types of baits and lures . . But I think there is only one true best bait. A Big . . Fat . . Juicy Worm and if the worm doesn't catch 'em nothing else will.

Let's say you're going out to harvest your own nightcrawlers. You have a good grassy area, a good heavy spring rain was falling throughout the day, and the temperature is about sixty to seventy degrees. It's a nice evening. It's dark. Before this time the worms have been dormant all winter. Now they're ready to come up to feed and to mate, and this is when you harvest them.

A lot of people do this wrong. They'll go out and harvest the nightcrawlers, and take them to an area where they're going to hold them. But they have no place to put them except in the small container they use for fishing, and that's overpopulated. It lacks proper bedding, and the nightcrawlers are already overheating. In the next couple of days, they'll die.

So let's start right from the beginning, before you even think of getting your nightcrawlers. I think the proper time to get ready for this is during winter, when you have a little extra time. Buy some material and build yourself a master worm holding box. The master container that I recommend is designed for five hundred to one thousand nightcrawlers. If you want to hold more crawlers, you'll have to build a larger container.

When the box is built, put about an inch of alfalfa hay or some burlap bags on the bottom. The reason for this is to obtain better drainage and also to keep the bedding from adhering to the bottom of the box. The alfalfa hay will also act as food for the worms.

The box should be held off the ground six inches or more. It depends on how convenient you want to make it, how handy to get at. You could put legs on the box, so that you could stand over it without bending too much. Or you could set the box on bricks or anything high enough to allow the water to drain.

The diagram below shows how you can cut all the necessary pieces for this box from two 12' boards:

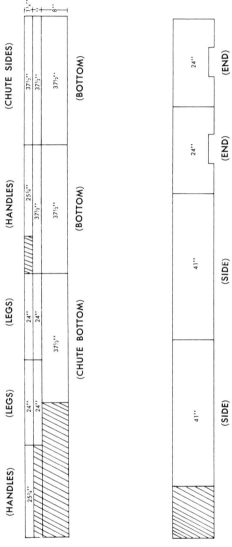

Figure 1   Cutting Instructions

Building A Master Worm Box

You can build a master worm container from two 12' lengths of fir or pine 1" x 12" boards. If you're handy with a hammer, saw, and drill, you'll have no problem at all building it. You'll need the following materials:

LIST OF MATERIALS

2 - 12' lengths 1x12 (fir or pine)
8 - 2" long 3/16" dia. stove bolts
8 washers
8 nuts
approx. 100 8-penny common nails
reg. door screen — 36" width —
   1 1/4 running ft.
   (aluminum or fiberglass)
approx. 100 1/2" wood staples

INSTRUCTIONS FOR ASSEMBLING: After boards have been cut as illustrated, assemble in the following way:

A. Construct bottom of box first. Nail sides of ventilation chute to bottom boards. *Then* nail *bottom* of ventilation chute to the sides of ventilation chute. Next nail center support board of ventilation chute to bottom of ventilation chute.

B. Drill 3/4" vent holes as shown. On sides and end boards, center holes 2" from bottom of box. Cover with screen. Use 1/2" staples.

C. Nail *ends* of box on. Note that floor of box is recessed 1/4". Note also that ventilation chute is *longer* than bottom of box. This is to facilitate end fitting over chute. Toe-nail chute to ends.

D. Nail on sides (line up top edge of box). Center sides to allow legs to be placed on ends. Legs may be nailed to ends and bolted to sides.

E. Attach handles by nailing to legs.

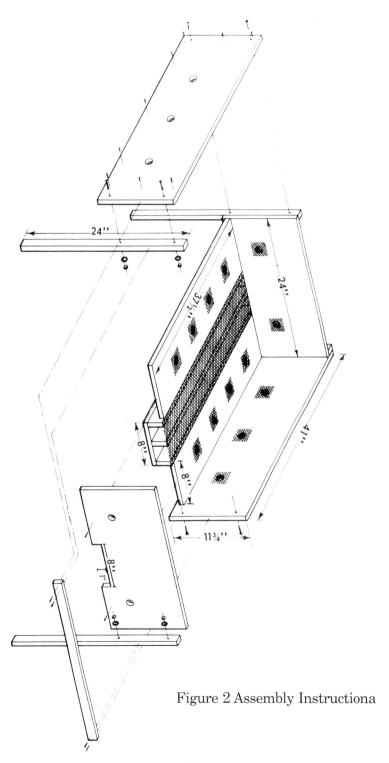

Figure 2 Assembly Instructiona

1

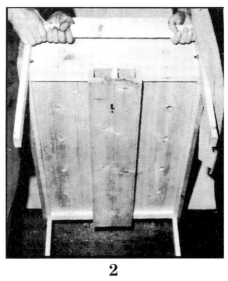

2

3

Figure 3 Views of Finished Worm Box

Picture number one shows completed box. Number two photo shows ventilation channel at bottom of box. (As natural ground is ventilated by insects — ants, etc. — channel imitates natural conditions.) The third photo shows worms clinging to the suggested natural sod covering.

Bedding

Now that you have the master container, what are you going to use for bedding? At the Magic Worm Laboratories I spent hours, many hours, in working with all types of bedding on the market. I tested many beddings, followed instructions, and used them with various types of worms. And I came to the conclusion that the type of bedding I use to operate the Magic Worm Laboratories is best. It's called Magic Worm Bedding.

Good bedding must have the following:

1. It must be organic, a natural bedding to provide a natural home for the worms.
2. It must be able to absorb moisture. The bedding I use at Magic Laboratories will absorb about twenty times its own weight in water. This is very important because a worm's body consists of about eighty-five percent water. You must keep the worm moist at all times or the worm will not breathe through its skin and will die.
3. The bedding must be easy to work with. It should not be too dusty. It should not pack too easily.

All of these qualities are found in the Magic Worm Bedding, which has a two-fold use. First you can use it for worm bedding. Also, when it is worn out, instead of throwing it in the garbage can as you would some beddings, you can use it for potting soil. It makes some of the best potting soil in the world. All you need to do is mix it — one-third aged worm bedding and two-thirds soil. You can use it for just about anything that grows. You know, there are thousands and thousands of dollars spent for potting soil every year. With this bedding, you can raise worms and then raise flowers at no additional expense.

Good worm bedding is absorbent. It holds water like a sponge. It's healthy for the nightcrawler and stimulates growth. You see, a nightcrawler breathes through its skin. That skin must be kept moist at all time. If it dries out the worm cannot breathe and will die.

Watering

I suggest that you use lake water, river water, or rain water to add to your bedding. If the water is heavily chlorinated or contains fluoride, I suggest not using it, unless you have some way of eliminating the chlorine or fluoride. This may be done by setting the water in open pans for a number of hours or days, thus allowing the chlorine or fluoride to dissipate into the air. But I think it's best to use the natural water sources I've suggested.

When you add water to the bedding, stir it. Don't just dump the water in like a lot of people do, because the water will go right through the fiber and the worms will suffer. Magic Worm Bedding is made of millions of tiny fibers. It's spongy, and will hold the moisture once it has a chance to absorb it. If you just rush water through the bedding, it will go to the bottom of the container, and you will not have the proper amount of moisture.

So, add the water slowly and stir it with a stick or your hand. The bedding is very clean. It's not smelly or dirty. It's not dusty. Use one quart of water for each pound of worm bedding. If you should accidentally add too much water to your bedding, you can control it by adding fresh Magic Worm Bedding directly from the bag.

Too much water in the bedding will give your crawlers a whitish, washed-out appearance. They will come to the top of the bed, looking thin and weak. Too much water will also destroy the minerals which are very important for proper health and growth. When this happens, add more bedding to the container.

Then let this mixture sit for twenty-four hours or more. The longer you let it age, the better. Once you add moisture to the bedding, it does have a certain amount of heat. If you put worms into the bedding within a couple of hours, you're going to lose your worms. Cool the bedding down below fifty degrees before adding worms. Remember, ninety percent of the worms are lost because people do not follow instructions. They don't add the proper amount of water, they don't let the bedding age properly, and they don't cool it properly. If you put your nightcrawlers in sixty-five or seventy degree temperatures, you're going to lose you worms just as sure as I'm writing this book.

Controlling Temperature

Now you will say to yourself, "Where am I going to put this bedding? My basement is the only place I have." Well, I'll tell you something. I'd say eighty or ninety percent of the worms die because people keep them in their basements. Basements are not cool enough for nightcrawlers. You say it's cool. How cool? Have you tested its temperature? Have you taken a thermometer test of the bedding? I'll bet it's seventy or seventy-five in your basement

The number one killer of crawlers is improper temperature. I always keep a thermometer about three inches into the bedding and do not let the temperature go above fifty degrees.

in summer. And that's cool, when you're coming in from ninety degree weather. But it's not cool enough for nightcrawlers.

Here's what you can do. Buy yourself a simple, cheap thermometer. Put the thermometer about four or five inches into the bedding. Then read the temperature after a number of hours. If that temperature is below fifty degrees, your bedding is in a proper temperature range for holding nightcrawlers. If it's above fifty degrees, that's it!

How You Can Cool The Bedding

I would suggest that if you don't have a large refrigerator to hold your container, get some ice-packs or some ice cubes in plastic bags, and put them on top of your bedding. Or get some refreezable picnic ice units and put them on top of your bedding. Or use cans of ice. Anything that will cool the bedding, because they'll melt, and add too much moisture. Put the cubes in plastic bags, or cans, or plastic containers of any kind.

I travel the road with my nightcrawlers thousands and thousands of miles during the summer in hot weather, eighty and ninety degrees. I can leave home with a plastic container of ice cubes, get to my room at night, check my worms, and put some more ice on them. In the morning, when I leave the hotel or motel room and come home again, my nightcrawlers are in perfect condition. They're solid, firm, active. At home, I put them in their master container, a walk-in refrigerator with a temperature of around fifty degrees, and I hold them there until I go on my next trip. If I can do it, you can do it!

Amount of Bedding

Another mistake that a lot of people make, is putting bedding too deep in their containers. I suggest that you have anywhere from four to six inches of bedding in your master worm box. The deeper the bedding, the more money it costs you, and the more chance for producing too much acid in your bedding. You'll have better aeration and less acid in your bed, if you have a shallow layer of worm bedding in your container. That's why I suggest a depth of between four and six inches.

Feeding Time

To feed your worms, use Magic Worm Food. This is especially made for earthworms, has thirty-two ingredients, and is high in protein, minerals, vitamins, and carbohydrates. Put two rows of worm food, about one-fourth inch wide and one-eighth inch deep, along each side of your holding box. (I place the rows about four inches in from each side.) You can either put the food on top of the bedding or submerge it an inch or so. Do not mix — and I repeat — do not mix the food down into the bedding. Down there the food will generate heat and develop too much acid. This can kill worms.

The bedding must have a certain amount of acid in order to prepare the food for the worms, and to help the worms digest the food. Worms do not manufacture acid like other animals to aid in digesting food. They must depend on the acid present in their food bedding to dissolve the food in their stomachs.

If you're having a problem with over-acid conditions in your worm beds, mix fresh worm bedding with your old bedding. This will make the old bedding drier, and aid in allowing gases to escape from the old bedding.

Be sure you have food on top of that bedding at all times. Check it daily. If the food is gone, add more. I also suggest that once every three or four weeks, you put a little brown sugar on top of your bed. Or a little dried, pow-

The Worm Czar who at one time was a feed technician has tried various feeds to produce a better conditioned worm. Besides feeds he has fed worms beer, Pepsi, volcanic ash from St. Helen's volcano, cake, powdered sugar, molasses and chocolate cake.

dered milk every three weeks. This gives them a change in diet, and they really enjoy it.

I often feed Herman the Worm a little ice cream once in awhile. It's very good for him. Your worms would like it too!

The more your crawlers eat, the faster they grow. Remember, it takes many pounds of feed to produce a pound of pork or a pound of beef. And it takes many ounces of feed to produce an ounce of worms. The sooner that worm gets the food into its body, the faster it's going to grow, and the quicker it's going to be in condition.

Next you ought to cover the container with a damp towel or burlap bag. I like to use burlap because it's very inexpensive, and the worms like burlap. You can use any piece of cloth, as far as that goes. Newspapers are all right, too, but they dry out very fast. It's hard to control the moisture with them. So I use burlap bags. By all means, if you get these bags from a feed mill or some feed dealer, wash them out thoroughly. Many feed bags have traces of medications in them. These chemicals can kill the worms. I know of a case where a gentleman lost thousands of worms in this particular way, by using unwashed burlap bags. So, wash them out.

The burlap should be kept moist at all times. Sprinkle it very, very freely with water. You might also place a sheet of dry cardboard on top of the burlap bag to hold out the light. Remember, nightcrawlers are very sensitive to light. Even a good, strong moonlight may send them down. So keep cardboard on top of the burlap bag. This will allow the nightcrawlers to be in a darkened area and to come up not only in the evening, but also during the day.

Keeping Your Worms Happy

To hold nightcrawlers in a master container for the summer, I would suggest a grass sod top instead of burlap. In other words, instead of putting wet burlap on top of your container, go out in your back yard and get some grass sod to use (similar to the thickness that you'd buy at a greenhouse). Be very careful that it isn't chemically treated. If you get it rolled up from a greenhouse, and you're sure it's free of chemicals or sprays, that's fine. But I know of a man who lost many, many nightcrawlers because he bought chemically treated grass sod from a greenhouse. He put it on top of his bedding, watered it like I suggested, had the food under the sod, and he lost thousands of worms. The night before, the greenhouse had sprayed the sod with dandelion killer. So I would suggest that you get your own sod. Place the food underneath the sod. Then roll the sod back every now and then to see if the food is gone. You'll find worms underneath there!

The worms will think they're out in their natural habitat. They'll even mate on top of the grass. And the grass will grow, because you're going to water the sod just like you would add moisture to your burlap covering. When the grass grows four or five inches, cut it. Let it lie right back down on the sod. Later on, the worms will use it as food. If you've had trouble holding nightcrawlers before, this method might be just what you need. The worms will be right on top. That's where they should be, so they're very easy to harvest from your master containers.

Sometimes you may see mold on top of your bedding or on top of the food. Now don't let this mold scare you. It will not kill your worms. As a matter of fact, I think it's good for them.

So, now you know how to construct a master container, how to mix the bedding, how to add worm food, and what to use for a cover. By all means, remember to control the temperature of your new container. This will keep your worms happy.

Worms, just hatched, looking like a number fifty white thread.

Chapter 4

MANAGEMENT OF WORMS DURING THE HOLDING PERIOD

Big Fat Juicy Nightcrawlers, like these, will catch fish.

We have millions of people on Social Security who are interested in a inexpensive hobby . . here is where a worm project comes into the picture.

During the time when you're holding the worms for the summer and for the balance of the season, you'll have to go to where your worms are kept and read the temperature, check the food and see that the proper moisture is in the bedding every day. You want the worms to keep in tip-top condition. If they are consuming food and moving around on top of or down in the bedding, they're in good condition. You can also tell their condition by their looks and texture. If they feel firm and moist, they're in number one condition. If the bedding happens to be sticking to the worms, you have too little moisture in it. As I mentioned before, the worm's body must be kept moist at all times or it will die, because worms breathe through their skin. Remove any sick, damaged worms you find as you check them.

If you find a number of small flies in your bedding, it's due to overfeeding. You don't want this. To eliminate the flies, feed less but more often, and cover the food with about an inch of bedding. Again, by all means, do not mix the food down into the bedding.

If your bedding is starting to get worn out, you'll say, "Well, George, how long will the bedding last, anyway?" That's a good question. Many people ask me that. A good rule of thumb is, if the bedding begins to look like dirt, it's humus. If it's sticky and not fresh in color, it's worn out and time to replace it. If you don't remove it, you're going to have too much acid in that bedding, and you're going to kill your worms.

You do not have to remove all of the bedding. It's quite a chore. I suggest that you take the cardboard cover off, and the burlap bag, and put a strong light above the bedding. This light will send the worms down. As they go down, you can remove the top of the bedding. Replace this with new, aged, cool bedding and mix it together. That will take care of you for some time.

If you notice that your worms are sick or dying, you can add some Antibiotic Terramycin Soluble Powder in with it. This can be purchased at most feed stores. Add one teaspoonful to eight ounces of worm food, then moisten it and scatter it on top of your worm bed. I have even lightly scattered the antibiotic alone on top of the bed — with excellent results

And remember, you don't have to throw your discarded bedding in a garbage can. You have humus here. (See Page 59 for mixing directions.) Give this potting soil to your friends who are gardeners, for flowers, trees, shrubs, or grass. Or package it and sell it on the open market. The demand is tremendous.

Now, to increase the worm's consumption of food, moisten it just a little bit. I'll tell you why. Worms cannot eat anything that's dry. They must moisten it first, and they usually do that with their own saliva. That takes time. So if you moisten the food, it will increase consumption. And the more food the worms eat, the better condition they'll be in.

Do not stir the bedding in your master container with your hands more often than necessary. Every time you do that, you disturb the nightcrawlers' habits, and they have to settle down to housekeeping again. If your bedding gets hard or firm, then I would suggest stirring the bedding or turning it over once every four or five weeks. I try to stay away from it altogether. The only time I dig into the bedding is when I change it to add new bedding.

Mixing Worms

Another question that I have been asked a number of times is, "Can you mix the smaller redworms with the nightcrawlers?" Well, I have done this and I cannot see where one worm will kill the other, like some people think. But I like to see the nightcrawlers held in a separate container. The nightcrawler is a larger animal and has different eating habits and different living habits. Also, the redworm is a more prolific breeder, and can multiply so fast that they can crowd some of the other worms out. If you want to hold the worms for a very short time, yes, mix the worms together.

Cautions

In summary, then, watch the bedding during the summer season. If you use too much or too little water, if you maintain an improper temperature, if you use improper worm food, if you handle the worms improperly, if you allow an over-acid condition to develop, you could lose your worms. Remember also to remove injured crawlers. And do not allow the bed to get overcrowded.

Just one more caution. Look out for rats, mice, moles, and ants. If they get into the worm bed, they'll put you out of business. To control rodents, I suggest you use traps. Poisons might get into your worm bedding and kill the worms. To control ants, dust chloradine powder around each worm box leg.

Above all else, check for proper temperature daily. Overheating is the biggest killer of nightcrawlers. If you follow these instructions, I think you'll be successful at holding these nightcrawlers the entire season.

Worm castings, true organic fertilizer and soil conditioner rich in Nitrogen, Potash and Phosphorus for everything that grows . . can bring in extra income for the worm grower. Worm castings (fertilizer) is selling for as much as $2.00 per pound in some markets.

George Sroda, the Nation's Worm Czar studies nightcrawlers, in his worm laboratory at Amherst Jct., WI 54407.

Chapter 5

TRANSPORTING WORMS DURING FISHING TRIPS

I think ninety percent of the nightcrawlers and other worms are injured or lost due to improper management while transporting them. In the summer, some fishermen travel for hundreds and hundreds of miles. They leave home with healthy, conditioned worms. They put them into a closed container that is too crowded and too hot. When they get to where they're going, they discover their worms are sick and dying, and blame it on everything else but their management program. So, let me explain the proper way to transport worms.

First of all, take along only as many worms as you'll need. Handle the worms as little as possible. Your hands leave a human scent on the worms. Wash your hands clean of any insecticides, detergents, sprays, or chemicals that you might have touched prior to handling your nightcrawlers. Many worms are dead before they get on the hook and into the water. It is a fact that many fish will bypass a sick, dying worm. Therefore, I suggest that you use a proper container. Too many people will buy a styrofoam container because it's cheap. They put their worms in it and put the cover on. The cover closes off the oxygen, and the worms die from lack of air. And these people wonder why they lose their worms! A tight cover allows too much heat to generate inside. Once the worms are overheated, there's nothing that can save them.

    I recommend an aerated bucket that I have developed in my laboratories. It has five large red plugs, two on each side and one on top, and these plugs are peppered with tiny breathing holes. These holes aerate the bedding. Someone might say, "Well, I'll lose the moisture in the container that way." Sure, you'll lose a certain amount of moisture, but it's the evaporation of moisture in a bedding that cools it. You can replace lost moisture by placing a small, damp cloth on top of the bedding. Or you can use an ice-pack on top of that cloth to keep the worms cool and moist. Don't worry about losing that moisture.

On the Road

By all means, put some cooling agent on top of the bedding and on top of the cloth for traveling. Use either ice cubes in plastic bags or plastic containers, or refreezable picnic units. And have a thermometer in the container also, so that you can check the temperature every time you stop. If it's a long trip, you might want to pull over at a roadside from time to time to check the temperature of your container. If it's getting above fifty, you'd better get some more ice cubes or some other agent to cool it down.

The next question is, where are you going to set your container in the car? Too many people put their worm containers in the trunks of their cars. That's the worst place! And by all means, do not put them on the floorboard of your car, above the exhaust pipe. Regardless of what cooling agent you use, exhaust has a lot of heat, and it can overheat the container and kill the worms. Put the container on the back seat of your car, out of the sunlight. If you have an air-conditioner, put the container of worms near that. That'll keep them in very good condition. Then, when you reach your destination, your worms will be as healthy and as lively as when you left home.

Another thing to consider is vibration. Vibration has killed a lot of worms. A car travels very fast and often over rough roads. The vibration of the car will shake the worms, causing uneven distribution, and making them nervous. (Worms have a well-developed nervous system.) By the time you get to your destination, vibration can cause a lot of damage. I suggest you set your container on the seat, which will absorb some of the vibration, or on an inner-tube, pillow, or anything that will reduce the shaking while traveling. You want the best, healthy, conditioned worms you can have when you go on a fishing trip. You deserve it. After all, it's going to take the best worm to get the best fish.

Out Fishing

Now you could have the nicest worms when you leave home, control everything very well during transportation, but when you go out in the boat, the worms will be in the hot sun. Unless you have something in the container to keep it cool, you're going to overheat your worms. Once the worm is overheated and dying, the fish will not take it. Every good fisherman knows this. So again, have the proper container out in the boat. And take along only the number of worms you plan to use that afternoon or morning. Use a cooling agent there too, just like in the car.

To store nightcrawlers, away from home at your summer cottage if you do not have refrigeration, dig a hole underneath your cottage or someplace where your worm box will be out of the sun and rain. Dig the hole six inches deeper than your worm box to help keep it cool. Here you must have a solid cover to protect the worms from coons, reptiles, and rodents. Follow the instructions for holding crawlers indoors, as to proper moisture, coverings, and feeding. Be sure to maintain your bedding temperature at fifty degrees or cooler at all times. Then you'll have happy, healthy worms when you put them on the hook. They'll wiggle themselves crazy, attracting many fish. But they will not have any action if they're dying from being over-heated.

Baiting Your Hook

If you have followed my suggestions in regard to harvesting and general care of nightcrawlers, you will have the liveliest bait known to man. Now don't spoil everything at this time by baiting your hook improperly. Bait your hook so the crawler will act and look natural. Hook your crawler once through the head, and not doubled and cramped like a pretzel. Or if you wish to use a weedless hook, hook the crawler first through the head and then into the collar.

Ice Fishing

Now let me suggest just a few things about how to control worms while out on the ice in winter. You must prevent them from freezing. Use the same type of container, and the same kind of moist bedding. But in the winter you'll need a heating agent to prevent freezing. I have discovered that if you get pocket hand warmers, which can be purchased at any sport shop, and put one or two on top of your bedding, this will control the temperature. You can use these with your permanent worm bed as well. Then you can fish all winter long with your worms.

Like a scientist, George studies worms and various types of worm bedding for healthy worms.

I use a cooling agent to control the temperature of my worms.

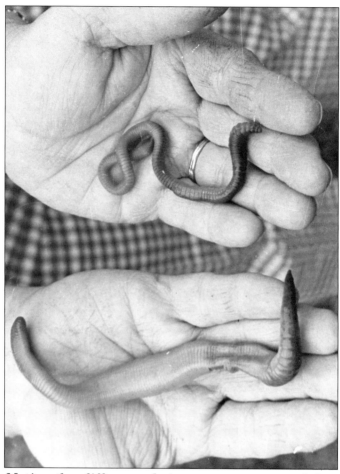

Notice the difference between an average nightcrawler and a conditioned crawler that no fish can resist. This type of conditioned crawler will wiggle like crazy on the hook, attracting fish.

These worm boxes, with worm bedding and worms can be used as a composting unit producing Organic Soil Conditioner.

Worms turn, twist and jump. Under Sroda's program a worm like this will attract fish.

Chapter 6
THE REDWORM

So far in my book, I have discussed the nightcrawler more than anything else. Now I would like to consider the redworm, often referred to as the hybrid worm. The redworm is much smaller than a nightcrawler. Here at the laboratories, I have raised them up to five or six inches in length. The redworm is a tough worm, capable of living a long time in water. The two most common types of redworms are the ones with yellow stripes at each segment, and all-solid redworm. Both are good for indoor breeding, and both provide excellent fishing bait.

The redworm is quite easy to raise. It can stand a much higher temperature than the nightcrawler. I have found that the best mating and hatching temperature is about seventy to seventy-five degrees. They can even tolerate up to eighty degrees at times. With eighty degree temperature, I usually have a circulating fan blowing at one end of the row of redworm boxes.

By having the air circulating, I can get by with such temperatures.

Breeding Redworms

Redworms lay worm capsules every seven to ten days which will hatch in about three to four weeks if the management is proper. The offspring will be ready to mate and produce young in about three to four months. Each capsule they produce will have anywhere from two to twenty baby worms.

You can begin a redworm business with one fertile worm, or one fertile egg capsule. But to speed up the process, place about two or three hundred capsules per cubic foot of worm bedding. Cover the capsules slightly with bedding. When they hatch and begin to grow, divide them into more beds to avoid overpopulating a single bed. So prolific is the small redworm, that one thousand multiply to one million in a year. In two years there would be around two billion.

The feeding and care of redworms, outside of temperature, is quite similar to that of nightcrawlers. Here at the laboratories, I feed the redworm the same food, use the same bedding, and follow the same management program as I use for crawlers. You can use a worm holding box similar to the type discussed for nightcrawlers. (See Pages 52-57.) Use no more than five or six inches of bedding as the redworm does most of its work near the top of the bedding. And of course with shallow bedding, you will have less chance for over-acid conditions, because there is more aeration. Bedding that packs and has more depth to it will increase the chance for over-acid conditions.

Getting Rid of Flies and Mites

Now, because of the higher temperature these redworms tolerate, you may have a fly problem. You might also notice tiny white worms, or white or red mites. These insects are harmless, although they will consume some of the worm food. They will not kill your worms. If you have a large number of these red mites, probably your temperature is too high. Mites like high temperatures. They live on the acids manufactured by bacteria in the bedding. These insects, or mites, are also found in the soil, the worm's natural habitat. And they find their way into worm beds too.

I have tried different ways to eliminate these creatures with no success. It's difficult to find a chemical that is safe for the worms. But I have used the following method to help hold down

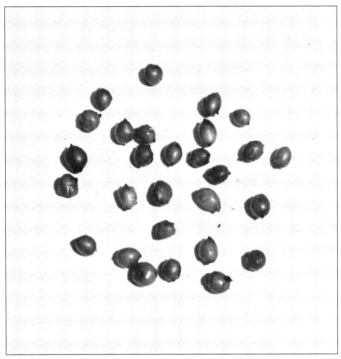

Nightcrawler eggs or capsules. Crawlers start producing egg capsules in their second year or when the collar appears. Capsules will hatch within sixty of ninety days. Worm capsules can be held dormant for months in low temperatures around forty degrees.

their population, and it works very well. Take an aluminum pie plate or a piece of cardboard soaked in water (even a pancake, or a melon rind will work well) and place it on top of your bedding. In a few days, it will be loaded with tiny mites. When this happens, take it out, shake off the mites, and replace it. It helps to put a little moist worm food on the cardboard or in the aluminum pie plate. Being moist, it'll start to sour, and bacteria will form, attracting the mites. I have used this method for years and have found that over a period of time, the mites disappear.

In order to control the fly problem, I suggest that you cover the food with about an inch of bedding. There's nothing on the market that will kill the flies and not hurt the worms. Therefore, it seems best just to cover the food.

Sometimes the feed is heavily contaminated with mites. So, I remove the feed and dispose of it. I might waste a little bit of feed, but I think it's worth it.

Sex Life of Redworms

The redworm is similar to the nightcrawler at mating. It merely slips its head under the band or collar of another worm. They are all the same sex but mate with another worm to produce young.

The worm capsules can be held for great lengths of time by holding them dormant. Let them dry out and keep the capsules in low temperatures, about thirty-eight to forty degrees. Later you can take the capsules that have been held dormant, and place them in seventy to seventy-five degree temperatures with adequate moisture, and they will hatch.

It will take more than a year to get a good crop of redworms by holding the best breeders back for reproduction. Remember, if your worm bed becomes overpopulated, you can take half of the hatch box bedding and put it into a new box. Then you will have two boxes with half of the old bedding in each of them. Add fresh bedding to the two boxes that are half-filled and mix this, and you're back in business again. If the bed becomes too crowded, redworms will not breed well. Watch the moisture content too. If the bed is too wet, the worms will not do well. If it's too dry, they will die.

Redworms like to breed and settle in the center of the pit. I don't like to disturb this area any more than necessary.

The redworm is an easy worm to raise, and like the nightcrawler, it is invaluable for soil conditioning.

Mating time. Nightcrawlers mate heaviest in the spring. As you can see, the crawler acting as a male will bury its head in the other worm's collar, gluing the two together with a slimy solution while exchanging sperm. After releasing itself, it will act as a female, producing egg capsules. Although the crawler is considered a hermaphrodite (male-female), it must mate with another worm to produce young.

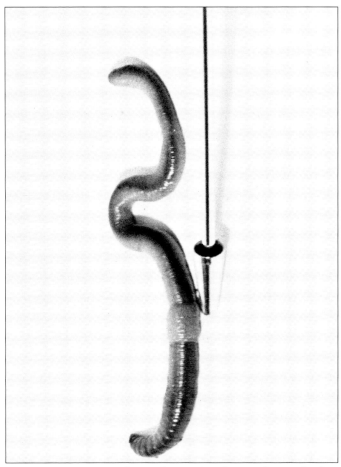

The pointer touches the raised portion of the worm's body called the clitellum or girdle. Two worms wiggle up to each other's girdle to mate. When they mate, the slime ring produced by this girdle slides forward and over the worm's head, producing a worm capsule. This will hatch, producing one or more baby worms.

Chapter 7

WORMS AND SOIL BUILDING

Earthworms play an important role in soil building. In the process of digestion, the worms extract food value from the food into their bodies. They use the food for strength, growth, and health. What they expel is in the form of castings, or humus. This humus is considered one of the best potting soils in the nation. It is truly organic. It usually has about five times more nitrogen, six or seven times more phosphorus, and ten or eleven times more potash than most potting soils. Earthworms are very important to all you gardeners.

Gardening is a popular hobby. Some people garden to help meet the rising cost of food. Others enjoy the exercise in the out-of-doors. And who doesn't enjoy beautiful, healthy, and colorful plants to cheer up the home inside? I tell you, both outdoor and indoor gardening can be done more successfully with the right

Earthworms work up the hard soil like a plow, allowing water and air to reach the plant roots. Worms produce organic fertilizer called castings or humus. All living plants love these castings. Worms help produce healthy plants.

kind of soil, with organic matter, and with good management. Because of the high cost of commercial fertilizer, more and more people are turning to organic gardening, and here is where the worm enters the picture.

For better outdoor gardens, greener grass, sharper colors, and less work for you, build the worm population in your soil by feeding them a special food, high in protein, minerals, vitamins, and antibiotics. Coffee grounds are not your answer. Neither is ground corn meal. Use the right kind of worm food, and the worms will begin to thrive in your soil.

Why earthworms in your soil? First, they will aerate your soil with their tunnels, allowing water to penetrate to the plant roots. Second, they will supply organic humus with their castings, and thus feed plant roots. And third, they will build up your soil.

Your yard can be the envy of the neighborhood. All you have to do is build up the earthworm population in your soil by adding redworms. Now, I suggest redworms because they do most of the work near the root system. The nightcrawler penetrates further down into the ground.

As I can see, there is a shortage of good healthy conditioned active red worms, easy to hold with very little losses.

How To Plant Worms In the Soil

Plant the worms in colonies. Place about two hundred in a hole twelve inches across and twelve inches deep. Next, add some worm bedding and some worm food to each hole. Place the holes about ten feet apart and keep watering them. Then watch the worms go to work. I think the earthworm is Nature's best friend!

If you're growing fruit trees and shrubs, dig several holes ten or twelve inches deep, and about ten or twelve inches wide, out near the branches' ends. Follow the same directions as you would for planting worms in grass gardens.

Soil that is well-populated with earthworms will be richer than most of the best topsoils. A good worm population in your soil will run anywhere from 300,000 to 2,000,000 worms per acre. Some soil may have as many as 6,000,000 worms per acre!

Earthworms in our soil also make good soil analysts. You can judge the quality and value of your soil by the number of earthworms and worm castings. If you have a heavy population of earthworms and a great number of castings, you know you have good topsoil.

"Earthworms for Monitoring Soils"

According to a science report, University of Wisconsin, Madison's, Philip Helmke and Paula Schomber, soil scientists in the UW college of Agriculture and Life Sciences, have successfully used earthworms in laboratory studies to monitor several metallic elements in soils. Earthworms provide a cheap and accurate way of doing this . . It was found that heavy metal pollution of soils, can reduce plant growth and can pose a health hazard by entering the food chains through plants or soil organisms.

Tests have shown that mulched plants produced more and have greater resistance to disease than those without mulch.

All life on this earth depends upon the productiveness of the soil.

Fertilizer tends to build up the salts and acids in the soil.

By feeding earthworms, one directly feeds plants that grow in the soil.

Good Worms Make Good Gardeners

I think that vegetables grown in this type of soil, heavily infested with earthworms as the main source of fertilizer, will contain more food value than those grown with commercial

Healthy soil produces large tasty vegetables shown here grown on soil well populated with earthworms on the test plot at the Worm Laboratory.

Earthworms produce the best organic fertilizer in the world. Here grass is grown in organic worm bedding with earthworms. No soil or other fertilizers were added.

In one corner of the worm laboratory work goes on with research adding worms to soil for numerous plants.

fertilizer. This soil will also produce healthier plants. In a number of areas, such as in the valley of the Nile, experiments have been going on for years, using earthworms as a main source of fertilizer. And great crops are grown there.

In tests with plants at my worm laboratory, I have found that plants grown with the proper number of earthworms, organic matter and worm food, have better growth, stronger stems, and sharper colors than the same types of plants grown in plain soil. In grass tests, the earthworm soil had more grass growth and greener color than the soil without earthworms.

I performed an experiment using an area of about 8,000 square feet with a newly built home on it. I brought in some black dirt and planted worms and food for these worms. I added some organic matter in the form of worm bedding and kept the plot well-watered. That yard had the nicest catch of grass that you could ever lay your eyes on. I used no commercial fertilizer whatsoever.

In another experiment I have three acres of grass that I haven't watered or fertilized for the last forty years. And I have a wonderful crop of grass. I let the worms do the work. All I do is see that they have worm food, which I

Note the difference in growth. The one contains just soil; the other contains soil, organic worm bedding and earthworms.

scatter on the ground so that they can come up at night and feed. Nature provides the water, but if a dry spell should occur, I see that the area is watered.

In one more experiment at the laboratories I took two cuttings from a coffee plant, both the same size. After rooting them in water, I put these cuttings into separate pots. In one pot I had soil alone. In the other pot, I had soil plus organic matter and worms. I kept the worms fed. Before long I could see a difference in growth. In a short time, the plant that had the worms in it outgrew the one with soil alone.

I believe that soil well-populated with earthworms and managed properly is healthier and has a stronger resistance to plant disease. Earthworms destroy the larvae of certain noxious insects in the soil.

Worms do a better job underground for growing things than some fertilizers can do on top. By adding commercial fertilizer to the topsoil, you can have a great amount of leaching or loss of fertilizer. But a good population of earthworms in the soil will help prevent commercial fertilizers from leaching and will also hold down water run-off after a rain.

Worm castings can also be used as top dressing for plants and lawns. Seedlings can be started in this mixture. In old lawns, remove some of the sod. Dig a hole under the sod, and add worm bedding, worm food and worms. Then water and sod again. When building a new lawn or planting seedlings, add the worms to the soil before the seed or seedling is planted.

It is better to plant younger worms or baby worms in the holes you dig than to use older breeding worms. This is because it is easier for the younger worms to adjust themselves to a new home.

As you can see, the worm is a valuable worker for building soil and helping to produce better, healthier plants. Just try some of my suggestions and see if they don't work for you.

"Attention Ladies . . Free Organic Potting Soil"

A SMALL COMPOSTING PROJECT FOR your plants while the man of the house has plenty of red worms for fishing.

All one needs is a Magic Worm Ranch which includes the worm bedding . . food and a temperature of about 65-70 degrees which is usually a basement temperature.

Mix the worm bedding according to directions . . add red worms which can be purchased by checking the classified section of any good sport magazine.

Add 500-1000 red worms on top of the bedding, add about a half cup of worm food on top of the bedding . . Remove cover and add a piece of plywood, one inch smaller than the inside of the worm box and you are in business.

Change the bedding when it is heavy and sticky. Now you have one of the best potting soils that you can mix, with two thirds soil and repeat the project. The fisherman will then have conditioned, happy worms for fishing. Happy Worms Make Happy Fishermen.

I do not sell the above worm box units . . they can be purchased at any large discount store.

Some of my best worm raising people are women.

Selling worm castings can make more money for the worm farmers than selling worms.

SAVE THE EARTHWORM AND WE WILL SAVE THE WORLD.

An inexpensive home compost unit . . no odor . . conditioned worms for the fishing enthusiast garden and organic potting soil for plants and gardens. A wonderful Hobby ...

An amazing tale ...
A WORM THAT PAINTS, A LIVING PAINTBRUSH.
A WORM THAT PLAYS BASKETBALL & MAKES BASKETS · · · ·

Herman is an unusual worm.

Herman the Worm is an unusual book.

It is a story of a sickly Canadian nightcrawler rescued by George Sroda, the Worm Czar, and nurtured back to health and to full growth, a majestic sixteen and a half inches, fully stretched.

The Life Story of TV Star & Celebrity Herman the

The story, told by Sroda as if in Herman's words, is factual, mixing information about nightcrawlers with accounts of Herman's adventures as a television star and media personality. Herman is likely the most widely publicized worm in America, having been written about in countless newspaper and magazine stories and appearing again and again on national and regional television and radio shows. What other worm has been seen on Johnny Carson's "The Tonight Show", "To Tell The Truth" and "The Mike Douglas Show" (six times!). What other worm has excited viewers with his pop art creations, as Sroda calls them?

ISBN-0-9604486-2-4

A WORM THAT TRAVELS AROUND IN A CHAUFFEURED CADILLAC · · · ·

"George Sroda (The Worm Czar) opened the door to the world of earth-worms for me. His knowledge, experience, and authority are unquestioned, and I am delighted that he has now written the story of his favorite—Herman."

Jerry Minnich, writer and publisher of countless articles, and Assistant Director of the University of Wisconsin Press

"When George does something, he does it right; this book on Herman the Worm is a great delight! I didn't think he could top his *Facts About Nightcrawlers* but he's done it again."

Anna Glidden, Poet
The Little Flower Knew
Palo Alto, California

GEORGE SRODA
Amherst Jct., WI 54407
Phone: 715-824-3868

Chapter 8
MERCHANDISING WORMS

I've included this chapter for those readers who might be interested in selling worms for profit. I have many years of experience in merchandising, and I think I can give you some good, practical advice.

First of all, sell only healthy worms. Don't try to get by with squeezing a few sick, weak, dying worms into your container. Have your worms conditioned, and be sure you have the correct count. You want the customer to come back over and over again.

Another idea to bear in mind is to have the proper bedding in the container. Don't use ordinary soil and think you're going to cheapen your sale. You will not get your customer back. Ordinary soil will dry out your worms, and your customer will complain about the quality of your product. Use the same type of bedding for selling worms in individual cartons as you have in your master worm container.

Don't Let Your Worms Lose Their Cool

Keep the containers at the right temperature, around fifty degrees or lower for nightcrawlers, until your customer calls for them. Then tell your customer to keep them cool also. Explain to him the importance of temperature control. Educate him on how to care for his worms after he has purchased them.

I suggest that you have a good, clean container. It must have air holes so the worms can breathe. Don't try to get by with some old, rusty tin cans. Get a good worm container. You can buy a number of them on the market. They're made for packaging worms and are labeled. "Bait." Make sure your container is clean. And it should be almost moisture-proof, because the bedding will contain enough moisture. You can get cups that have a waxed coating on them. And you can get your own labels printed at a print shop. You might want to have your name, address, and phone number printed on them. If you're selling quality worms, you should be proud that your name is on the container. It will advertise your product from one fisherman to another and tell them where they can buy nice, clean, attractively packaged worms.

There are over ten million women who are

fishing. I don't think these women want to pick up a dirty, soiled carton. Nor do they want to put their fingers into ordinary dirt and get it under their fingernails. If you have good, clean bedding, such as I recommend, these women will not have to worry about such things. Magic Worm Bedding is a pleasure to work with. It's soft and resilient. It has the feel of silk.

Selling Worms

You can sell your worms by the side of your road if you want to. Or put up a sign in front of your home. A number of people do this every spring. I know youngsters in my area who sell nightcrawlers every year. Some of these twelve and thirteen year olds make as much as $400 to $500 a year. It's also a good supplementary income for the elderly. Anyone can profit from selling worms. And it's a pleasure to sell to the fishermen, who are ideal customers. You can also run ads in your local newspaper, or even in national or state magazines that promote this product. Many people ship worms by parcel post, both to fishermen and to people in the business of raising worms.

Set a price that is competitive in your area, but don't worry too much about price. Fishermen will always pay more money for a

Selling earthworms is a big business. Properly packaged in clean, printed wax cartons, your healthy crawlers will bring a premium price.

good, active worm. You see, they will need fewer worms if the worms are of good quality, because they will last longer and catch more fish.

Remember, a good product establishes a good market.

At one time a reporter wrote a feature article on George Sroda "If George was selling door locks, every door would have five locks."

George studies worms for their body color. Fish go for live bait that's close in color to their natural foods. George's worms aren't black, just darker than most worms, closer to natural fish food colors.

Nightcrawlers, so well fed they look like miniature snakes. That's the world of Sroda, a worm expert from the heart of fishing country in Central Wisconsin, U.S.A.

CONCLUSION

I hope you have enjoyed reading this book as much as I have enjoyed writing it for you. There is much more work to be done in this field, and I expect to continue doing some of it.

I always remember the day I was walking alongside the famous Tomorrow River trout stream near my hometown in central Wisconsin. I watched fishermen pull sick, weak, dying worms out of cans of dry soil, and expect to catch fish. This prompted me to give up turkey farming and to set up the Magic Worm Laboratory, the largest privately-owned worm laboratory in the nation, and to do research on worms, especially the night-crawler. I don't miss turkeys, nor their gobble, for I have a new sound, the silent sound of the contented worm wiggling in its magic bed.

I can also remember from my earlier days, how I would stop my car alongside the road on a trip, run into the field of a farmer who was plowing, and gather worms for my experiments; and then how I accidentally scared a waitress by pulling a handkerchief from my pocket that held those freshly picked worms and scattering them all over the table. How in the early days my wife forced me and my worms from the kitchen, after they had escaped from their carton in the refrigerator and entered her salad. Yes, all these memories come back to me as I am writing this book.

If you're at all like me, you may wonder what makes the wild flowers look so sharp, so healthy, so strong, why the grass is so green, and why the forests stand so proud without human help. I think I've found part of the answer. I hope you will too.

I am the most widely publicized nightcrawler in America. I have been written about in countless newspapers and magazines stories, and appeared again and again on national television and radio shows. What other worm has excited viewers with pop art creations and entertainment?

George Sroda, the Czar of the Underworld, rules two million subjects who wouldn't think of rebellion.

IN A NUTSHELL

Tips for Fishermen:

1. Cool the bedding in your container to 50 degrees or lower before adding worms. Make sure there is ventilation.
2. Take only the number of worms you will need. Don't overcrowd your portable container.
3. Remove sick or injured worms from the container.
4. Take along a thermometer and check the temperature of the bedding frequently.
5. Cushion the container from vibration while transporting it.
6. Take a cooling agent with you.
7. Wash your hands before handling worms.
8. Hook the worms through their heads so that they look natural.

Tips for Worm Breeders:

1. Age the worm bedding at least 24 hours before adding worms.
2. Cool the bedding to 50 degrees or lower and maintain this temperature in your master container.
3. Remove sick or injured worms from your master container.
4. Place the worm food on top of the bedding or an inch below it. Do not wash the feed into the bedding and do not mix with the bedding.
5. Moisten the worm food to increase consumption.
6. Place a moist, washed burlap bag on top of the bedding and keep it moist.
7. Wash your hands before handling worms.
8. Do not over crowd the master worm box. The container I've described will accommodate 500 to 1000 nightcrawlers or twice as many smaller worms.

Dear Anglie

Q. Although I have heard you should not keep worms in too deep of a container even though it is filled with bedding, I still don't understand why?

A. With a deep, full container, you'll get "temperature build-up" within the bedding, causing danger to the worms. A shallow depth of four to six inches of bedding is recommended. We recommend the Magic Worm Farm, that features a series of ventilation plugs to aerate the bedding, regulating its heat to prevent worm suffocation and dehydration. Its height permits an ice container to be placed on top of bedding, keeping worms properly cooled while fishing.

Q. I don't have room in our refrigerator, so I am wondering if keeping the bait container with nightcrawlers in it on the basement floor is okay?

A. Although your basement may seem cool compared to other parts of the house, it may be warmer than the 40-50° range needed to keep the nightcrawler alive and healthy. One trick is to use two refreezable "ice packs". By alternating (one in the container, while the other is refreezing), you can keep your crawlers cool and comfortable in storage or in the boat. Note: Magic Worm Containers; either the Magic Worm Farm (10 qt. size) or the Magic Worm Ranch (14" x 20" x 7") were designed to fit easily into standard refrigerators.

Q. To save room, can we store redworms and nightcrawlers in the same container?

A. Because redworms (manure worms) and nightcrawlers require different temperature ranges, you will need separate containers. Worms that exist in similar conditions, such as garden worms and nightcrawlers, may be kept in the same container.

Q. How can I be sure that the bedding is receiving the proper amount of moisture?

A. First, cover the bedding with newspaper, burlap or any piece of cloth. Second, check the covering at regular intervals and set a watering schedule. By keeping this covering moist, the bedding will retain its proper moisture.

Q. Does worm bedding last forever? How can I tell if I need to replace it and what do I do with the old bedding?

A. Unfortunately, worm bedding does not last forever. When the bedding turns black, it is full of worm castings and needs to be replaced. Instead of throwing it out like paper beddings, you can use this natural humus as an excellent potting soil.

Q. I am an avid fisherman. I fish weekends with my entire family and during the week with a neighbor. How many worms can I keep on hand in a Magic Worm Ranch?

A. Plenty. The scaled dimensions of the Magic Worm Ranch allows you to keep up to 600 nightcrawlers or 1200 smaller worms for several weeks or 200 nightcrawlers or 400 smaller worms for years. Sounds like you could use two Magic Worm Ranches.

Q. When moistening my Magic Worm Bedding, can I use regular tap water?

A. We recommend that it's best not to use chlorinated water, since chlorine may kill your nightcrawlers. Rain water, lake water or well water will do fine.

Additional Worm News

Harvesting earthworms is a big business in Canada where most of crawlers come from.

On a good moist damp warm night a good professional worm picker can pick up to 10,000 in one night making hundreds of dollars per night.

They are then loaded in large semi-trailer trucks delivered to us distributors and sold at retail to fishermen. One out of every three americans fish. We have over 40 million fishermen in us.

I spend a lot of time in my worm laboratory researching the worm for color, action, smell, to attract fish. I receive a great deal of mail from people that don't fish but are interested in raising worms as a hoppy and for their garden, lawn and flower beds.

I receive a lot of mail from kids, who are interest in Herman the Worm and starting out in the worm sales business some of these kids make as high as $500.00 in selling worms. Yes! Worms can save the day . . If baits & lures can't catch 'em a nightcrawler will.

A Good Active Condition Worm Will Catch Fish.

A stream, a fishing dock, and a line with a happy, conditioned worm on the hook. A bucket of worms so lively that they knock the cover off with joy.

Earthworms make an excellent high protein food for some types of fish.

For more information about the products mentioned in this book or about worms in general please contact me.

>George Sroda
"The Worm Czar"
Amherst Jct., Wisconsin 54407

George Sroda: Central Wisconsin's Unique National Celebrity and Promoter Extraordinaire

From the Waupaca Publishing Co.
By Dan Hansen Editor - Writer

At first glance it may be difficult for most people to picture George Sroda as a celebrity with national and even international stature. He doesn't have the face of a Robert Redford or possess the body of an Arnold Schwarzenegger. He's never starred in a major motion picture or his own television series, held high political office, or been involved with the Kennedys.

All this notwithstanding, George Sroda has received an enormous amount of exposure - through various print and electronic media - to people throughout the United States, Canada and several other countries of the world.

He's been written about in such diverse publications as *Outdoor Life, Sports Afield, Organic Gardening, Grit, TV Screen, Fishing Facts, The Wall Street Journal* among others. He's appeared as a guest on the Tonight Show with Johnny Carson, the Mike Douglas Show, To Tell The Truth, Amazing Animals, the David Letterman Show, P.M. Magazine, Real

People, National Public Radio, Outdoor Wisconsin and many more.

Sroda has proven to be a popular media favorite because he's enthusiastic and animated.

So how does a small town boy from Amherst Junction, Wisconsin, grow up to be a successful businessman, photographer, book author and television personality? As is the case with most people who've earned success (rather than winning the lottery or marrying money), George Sroda's achievements came about over a period of many years through a combination of several different factors. These include a strong desire to succeed, a deep commitment to family and the work ethic and the ability to sell just about anything to almost anybody.

"I've been a salesman all my life," declared Sroda in a recent interview. "I could sell anything, but I've got to be sold on the product, I've got to love it and I've got to understand it."

Born into a Stevens Point family during the second decade of the 20th Century, he began honing his salesmanship skills at age 10 when he sold newspapers for two pennies a copy. He learned the basics of merchandising behind the counter of his parents' general store in Amherst Junction. After graduating from high school, he managed the store, earning $30 a month. Following his marriage to Susan Rekoske of Amherst, he received a $20 a month raise.

Sroda has always had an inventive mind and a wide range of interests. His musical talent on the banjo led to the formation of his own band at age 21. At age 24, he was bitten by the photography bug when he got his first Kodak box camera the cost $1.98, and has been as avid photographer ever since. He hardly ever goes anywhere without at least two 35 MM cameras - one loaded with color film and the other with black-and-white film. He develops and prints black-and-white photos in his own basement darkroom. A longtime member of the Wisconsin Press Photographers Association, Sroda has had numerous photos published in various state-wide circulation newspapers. Many of his photos have been used to illustrate the annual Waupaca County Visitors Guide and other publications. Over the years he had taken tens of thousands of photos, and estimates his current inventory to feature at least 5,000 color slides and more than 50,000 black-and-white negatives.

Sroda's hobby is photography. He spends many hours photographing various subjects and producing the finish photos in his own darkroom. He puts on film behavor of worms and siphons this information to people interested in worm.

Newspaper reporters and photographers beat a path to the worm laboratory for late worm knowledge shooting a lot of pictures along the famous Tomorrow River near the worm laboratory.

Recording and photographing worms at the laboratory.

Sroda's one, brief flirtation with big city life came right after his high school graduation when he tried living in Milwaukee. "I thought it would be great to work in a big department store there, but after only a week I discovered this was not the place for me."

When he decided to settle permanently in the tiny hamlet of Amherst Junction, and it's a decision he's never regretted. "I've proved that a person doesn't have to live in New York, Chicago or San Francisco to be successful," he explained.

For several years Sroda successfully operated a feed mill in Amherst Junction. Much of his time was spent developing nutritional feed rations for cattle, poultry and hogs.

Always on the lookout for new opportunities to feather his nest, George Sroda left the feed mill and entered the turkey business. His critics laughed, but Sroda had the last and best laugh. At a time when most everyone else was selling turkeys for 19 cents a pound, he was marketing an organically raised, oven-dressed turkey that commanded 80 cents a pound.

This business also brought Sroda's considerable promotional talents to the fore. "I was the first to retail turkey by the piece, the first with turkey steaks. I sold everything including the "gobble." Many of the turkey by products were turned into high protein poultry feed. He also marketed the droppings as an organic fertilizer called "Turkey Peat."

The turkey feathers were dyed and made into beautiful feather flowers. Wrestler "Gorgeous George" purchased several purple ones to help accentuate his gaudy costumes. Sroda also recalls selling a large number of the colored flowers to showgirls from Hurley, who used them as props during their dance routines.

While the turkey business was successful, it was also very labor intensive, according to Sroda. "Preparing the birds required a lot of hand plucking and a lot of other work by many people," he explained. "I had a lot of good workers, but they were getting older, and it was hard to find new workers to take their place." Rather than allow the quality of his products to suffer, Sroda decided to curtail his operation and seek out a new challenge.

A routine photography session along the banks of the nearby Tomorrow River at Amherst in the spring of 1963 marked a monumental change in Sroda's life, launching him in a new career that would ultimately lead to fame and financial freedom.

"While I was out there taking pictures of

the fishermen, I noticed they were trying to catch fish with sick looking, dried-up, half-dead worms on their hooks," he recalled. "Right then and there, I knew what these fishermen needed was a healthy, conditioned, lively worm."

He decided to apply the research techniques he successfully utilized in developing nutritional feed rations for farm animals in an attempt to produced healthier, better conditioned worms and nightcrawlers. His first discovery was that very little work was being done at the time to develop a better nightcrawler; in fact, he had the whole field pretty much to himself.

Sroda began testing various ingredients in an effort to develop bedding and food products that would enhance the color and scent of the crawlers. He says worms need to smell good and have a rich, dark color in order to attract the fish. He found this research to be a real educational experience, learning that worms have 150 segments, 300 kidneys and five hearts, along with a great deal of other information. "There area three basic types of worms," he explains, "red worms, garden or leaf worms and the nightcrawler - the king of worms."

Out of his research came the Magic Worm Bedding Company of Amherst Junction that

In his worm laboratory, George Sroda studies nightcrawlers on an almost daily basis to help develop healthier earthworms.

manufactured and marketed worm bedding, worm food and a variety of containers from the "Magic Worm Hut to the Magic Worm Ranch." Sroda personally took to the road selling these products to sporting goods stores and discounts stores. "Selling worm bedding and worm food isn't as easy as selling ice cream on an August afternoon," he admits. Sroda found successful markets for his products and found receptive audiences for his sermons on the value of worms to society.

Probably Sroda's most fortunate discovery, however, was the 3 1/2-inch sickly worm he found in a shipment of Canadian nightcrawlers. As Sroda recounts the tale: "I took this weak, shriveled up, undernourished worm, and, using the proper food, bedding and care, turned him into a strapping 16 1/2-inch nightcrawler.

Sroda then christened this crawler "Herman the Worm," accompanied Sroda on numerous national television shows and has been the focus of many newspaper and magazine articles. Sroda has also discovered many of Herman's hidden talents, including painting, playing basketball and football.

By 1970, the worm bedding business was booming. Along with packaging the bedding and food for his own label, Sroda was also working with another company to develop national and international markets. About this time he came to the conclusion that he much preferred being out selling his products to retailers and preaching the gospel of earth worms to the general public than being the CEO of a growing company.

"I had always planned to stay within a certain part of Wisconsin," he related, "but the business started mushrooming to a point where I needed to hire a manager and office staff, and I didn't really want to do that, so I sold the company."

Selling his business gave him the freedom to travel and promote the benefits of earthworms, continue his research in his worm laboratory (which he say is the largest such privately owned facility in the world) and to share his extensive knowledge of earthworms by authoring two books on the subject.

His first book, *Facts About Nightcrawlers*, was a pretty straightforward account aimed at anglers, gardeners, worm breeders and worm sellers. The second, titled: *"The Life Story of TV Star and Celebrity Herman the Worm,"* is a more lighthearted offering, combining factual information about nightcrawlers with accounts of Herman's (and George's) adventures as a television star and media personality. Each title is available from George Sroda, Amherst Junction, WI 54407.

Television, however, probably has been the favorite forum for the world's foremost earthworm evangelist and his sidekick. George Sroda and Herman the Worm have appeared on more than 25 national TV shows. "Our first appearance was on 'What's My Line'," said Sroda. "The one most people remember was The Tonight Show with Johnny Carson. We were on for the full eight minutes. We've also been with David Letterman twice and on the Mike Douglas Show seven times."

Sroda also has some interesting tales about other stars he's met on these shows. "Some of these people come in with their agents, producers, band leaders, make up artists and hair dressers, and here I am with just a can of worms," he chuckles. Was he ever nervous about going on TV? "No, But, some of these other people were really afraid because they knew if they goofed up on the Tonight Show, they could be out of the business the next day. If I goofed up, I knew I could just come back to Amherst Junction and work with my worms."

Even into his eighth decade, George Sroda

remains busier than many men half his age. With the able assistance of his gracious wife, Susan, he still answers letters from people all over the world with questions about earthworms. "I want everyone to know that I don't sell worms," he stressed. "And I don't sell worm bedding any more, either, I just do research."

He turns down most interview requests that require travel these days. He has produced a video tape about himself and his work which he makes available to various media and other groups. Earlier this spring, he was featured in the prestigious *Wall Street Journal*. "Several weeks later, and I'm still getting calls and letters from that article," he declares in amazement. George Sroda and his pal Herman are also prominently featured in the National Fishing Hall of Fame in Hayward, Wisconsin.

Earlier this week, George called to inform me that he'll be a guest on National Public Radio Saturday, "I'll be live on 150 public radio stations from coast to coast for about 20 minutes," the Worm Czar proudly announced.

What does the future hold for George Sroda? "I've had a full and happy life; I'm not looking for anything new. I'd just like to continue working in my worm laboratory, trying

Sroda receives dozens of letters each week from people throughout the world who have questions about earthworms.

to help produce a better nightcrawler," was his thoughtful reply.

"I don't do it for the money, I do it because I love it and for the knowledge I gain by studying them almost every day. I haven't had a true vacation in about 30 years. A couple of years ago I was feeling a little upset and uneasy, so I went to my doctor for some tests. All he told me was that I was doing too much for my age; he said I should relax a little bit and go somewhere I'd like to be more than any place in the world. I said, 'OK, I'll do that tomorrow.' The next day I went to my worm laboratory, because that's where I wanted to be more that any other place in the world."

The Worm Czar proudly admires Herman The Worm, his star and the king of all worms.

TV Star Herman The Worm and his master George Sroda on display at The Fishing Hall of Fame, Hayward, Wisconsin. Over 150,000 people visit the museum each season.

World wide newspaper reporters feature The Worm Czar and his TV Star and celebrity Herman The Worm in their publications.

Color Photo: Marshfield News-Herald

George Sroda, author and Worm Czar, in deep meditation holding his TV Star and Celebrity HERMAN THE WORM with some of Herman's paintings in the background.

HERMAN THE WORM
He's not just a star, but is a Super Star, having appeared with some of the top celebrities in Hollywood.

HERMAN THE WORM
He is the only earthworm in the world who paints with his body, a living paint brush, creating modern abstract designs.